Mathias K. B. Lüdecke, Martin Budde, Oles Kit, Diana Reckien

Evaluating Climate Change Scenarios

From AOGCMs to Hyderabad

Emerging megacities
Dicussion Papers
Edited by Konrad Hagedorn, Christine Werthmann, Dimitrios Zikos, Ramesh Chennamaneni

Humboldt-Universität zu Berlin
Department of Agricultural Economics
Division of Resource Economics
Philippstr. 13, House 12
10115 Berlin

Tel.: +49 (0)30 2093 6305
Fax: +49 (0)30 2093 6497
www.agrar.hu-berlin.de/struktur/institute/wisola/fg/ress
www.sustainable-hyderabad.de

Contact: emerging.megacities@hu-berlin.de

The emerging megacities discussion papers are available at:
www.eh-verlag.de

ISSN print edition 2193-6927

Emerging megacities Discussion Papers are prepared by researchers working on topics in the realm of sustainable development in Megacities of Tomorrow, a research priority by the German Ministry of Education and Research (BMBF). The papers have been peer-reviewed by a board of external reviewers.
Views and opinions expressed do not necessarily represent those of the Division of Resource Economics.
Comments are highly welcome and should be sent directly to the authors.
We welcome contributions on any topics related to Megacities of Tomorrow. Further information on the submission procedure is given at:
www.sustainable-hyderabad.de/emerging-megacities

Lüdecke, Mathias K. B.; Budde, Martin; Kit, Oles; Reckien, Diana

Evaluating Climate Change Scenarios
From AOGCMs to Hyderabad

Emerging megacities Discussion Papers, Volume 5/2009

ISBN/EAN: 978-3-86741-817-1

First published in 2012 by Europaeischer Hochschulverlag GmbH & Co KG, Bremen, Germany.

© Europaeischer Hochschulverlag GmbH & Co KG, Fahrenheitstr. 1, D-28359 Bremen (www.eh-verlag.de). All rights reserved.

Cover: Photo "Metropolis", ferendus (flickr). Creative Commons License

No part of this publication may be reproduced or transmitted, in any form or by any means, electronic, mechanical, photocopying, recording or otherwise, or stored in any retrieval system of nay nature, without the written permission of the copyright holder and the publisher, application for which shall be made to the publisher.

EHV

Evaluating Climate Change Scenarios
From AOGCMs to Hyderabad

Mathias K. B. Lüdecke[*][†], *Martin Budde*[†], *Oles Kit*[†], *Diana Reckien*[†]

May 2009

Abstract

The paper develops adequate evaluation algorithms for applying Atmospheric - Oceanic General Circulation Model (AOGCM) simulated climate scenarios over Hyderabad, India. The AOGCM simulated results were generated for the IPCC AR4 process and considered for a high (A2) and a low (B1) global CO_2-emission scenario. The considered time slices are 1981–2000 (reference climate), 2046–2065 and 2081–2100. These model runs were evaluated to obtain projections of the four most impact-relevant climatic characteristics for Hyderabad: the frequency distribution of daily precipitation (important, e.g, for urban flooding), the total annual precipitation (e.g. for urban water supply), the probability and duration of heat waves (e.g. for human health) and the annual mean temperature (e.g. for urban agriculture).

Key words: *climate change, local climate projections, Hyderabad, heat waves, intense rain*

---

[*] Corresponding author. Email: luedeke@pik-potsdam.de
[†] Potsdam Institute for Climate Impact Research (PIK), Research Domain: Climate Impacts and Vulnerabilities, P.O. Box 60 12 03, 14412 Potsdam

1 Introduction

The aim of this paper is to identify the climate change signals which the city of Hyderabad and its peri-urban region have to expect. This includes the investigation of the sensitivity of these projections with respect to global greenhouse gas emission scenarios and the consideration of uncertainty of the projections for fixed emission scenarios.

The presently most advanced knowledge on global climate change projections is represented by the AOGCM comparison done for the 4th Assessment Report of the IPCC (Meehl et al. 2007). Here 22 different AOGCMs were driven by a standardised set of global emission scenarios (SRES, Nakićenović and Swart 2000) and generated contrastable results, amongst others for temperature and precipitation change.

The investigations in the first project year on climate change impact paths in Hyderabad (see Reckien et al. 2011) revealed four major climate variables with relevant impacts on urban functions (see also Satterthwaite et al. 2007, McMichael et al. 2003, Revi 2007):

1. The frequency distribution of daily precipitation in the urban area, in particular the frequency of heavy rain events which cause urban flooding.

2. The frequency and length of heat waves in the city, generating direct health and infrastructure impacts.

3. The total annual precipitation in the region of Hyderabad, influencing the water supply of the city and the success of peri-urban agriculture.

4. The mean annual temperature in the region of Hyderabad, influencing via the natural water balance the peri-urban agriculture and groundwater refill.

To scan the range of possible future developments of the driving force for climate change, the global greenhouse gas emissions, we choose in this paper the SRES-A2 scenario for a high-emission future and the SRES-B1 scenario for a significant global emission reduction path.

Tebaldi et al. (2006) tried to deduce projections for 10 different indices of extreme weather events from a subset of the IPCC AOGCMs and could show that the signs of observed significant trends in the 20th century were reasonably reproduced by the model set. Based on this model validation, they calculated projections for the extreme indices, again with respect to the sign of the trend. They did the analysis for an intermediate emission scenario (A1B). Two of their indices are related to the climate variables #1 and #2 chosen for this paper. For the end of the 21st century they identify an increasing trend of days with a precipitation larger than the current 95 %-quantile and an increase in the number of hot nights as well as in the duration of heat waves for a large region around Hyderabad.

In this paper, efforts are done on analysis towards the quantification of changes, two future time slices (2046–2065 and 2081–2100) and emission scenarios as well as the full frequency distribution of daily precipitation. This is possible because we do not perform a global study but concentrate on one "point" (Hyderabad) and the surrounding area which allows us to calibrate the model data with observational data. And it is necessary because the planned quantitative impact study for Hyderabad needs quantitative estimates of the projected climate change stimuli.

This aim poses the question of how to "downscale" the AOGCM-results onto small regions (compared to AOGCM grid sizes) like urban areas. Here two possibilities can be distinguished: the dynamical (e.g. Hollweg et al. 2008) and the statistical (e.g. Orlowsky et al. 2008) downscaling. The latter relies on statistical relations between observed weather station data and either model variables or observed large scale weather variables which can be potentially well reproduced by AOGCMs. The first relies on nesting fine-scale, regional circulation models into coarse-scale global AOGCM runs. This is a consistent and elegant approach, but two shortcomings make this way questionable for our purposes: first, the difference in outcomes of the AOGCMs is still considerable and the results generated by the nested regional models are to a large extent prescribed by dynamic boundary conditions given by the AOGCMs. So, for a proper uncertainty analysis (comparable to the state of the art set by IPCC-AR4) runs under different AOGCMs are mandatory. The second problem lies in model errors of the nested regional model itself. To get an idea on the implied uncertainties, a comparison of alternative dynamic regional models would be necessary.

For India, currently there exists one study performing a dynamical downscaling using the regional model PRECIS driven by a HadAM3H-run (Rupa Kumar et al. 2006). To illustrate the above argument we show changes in annual precipitation for central India as calculated by the different IPCC-AR4 AOGCMs in Figure 1– fixed to pre-given inflows and outflows of water vapour there is not much room left for the nested dynamic model to improve this – possibly wrong - precondition.

Taking all this into account, we developed a statistical downscaling procedure, adapted it to our specific task and describe it in detail in Section 2. A full fledged uncertainty analysis on the basis of all 22 IPCC-AR4 AOGCMs is part of ongoing work in a later stage

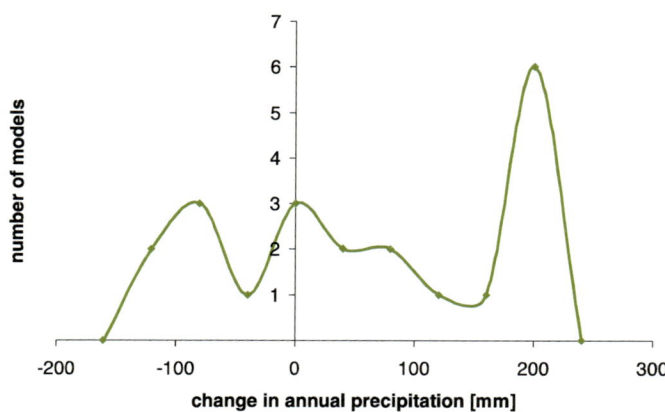

Figure 1: Frequency distribution of projected absolute changes in the amount of annual precipitation for the larger region around Hyderabad as calculated by 21 different AOGCMs. Time horizon: 2100; Emissions: A1B (intermediate emission scenario)

Source: own work, CERA data base

of the project. However, to get an idea on model induced uncertainty ranges we used already two AOGCMs for this study, which cover the moderately positive (Echam5) and the moderately negative (GFDL-CM2.1) part of Figure 1. In Section 3, we describe the data sources, their retrieval and preparation, Section 4 presents the resulting projections for the four Hyderabad relevant climate variables and Section 5 draws the conclusions and discusses direct policy implications.

2 Evaluation Algorithms

In the following, we describe the downscaling and error-correcting algorithms from AOGCM results to the local Hyderabad situation. The considered variables are the distribution of daily precipitation, the number of very hot nights per year and the duration of heat waves. For the areal variables annual precipitation and average temperature, we use the relative changes from the comparison of the AOGCM runs directly.

2.1 Evaluation Algorithm for Daily Precipitation

The downscaling algorithm for daily precipitation developed for this study is mainly a two step approach. In a first step the transformation matrix between the frequency distributions of daily areal and daily station data is determined. This is necessary because AOGCM results represent spatial averages over large areas (about 2° x 2°, "areal data") while the climate impact study concentrates on a comparatively small urban area ("point data"). The second, equally important step is the calibration of the present time (1981–2000) AOGCM results to the observed areal data. This calibration is then fixed and applied to the AOGCM projections for 2046–2065 and 2081–2100, thereby compensating for the errors in the reproduction of the observed climate and focusing the analysis on projected signals of climate change.

The necessary inputs for this algorithm are observation based time series for daily precipitation at a station P_t^S, and a larger area around the station (fitting the AOGCM grid cell), P_t^A. t denotes the number of the day. Furthermore, daily precipitation time-series modelled by an AOGCM, P_t^M. As a climate model does not generate concrete weather data, t is not a real date and comparison of these time series with the observed ones has to be based only on statistical properties. Here the discrete frequency distributions $\vec{f}^X$ of the daily precipitation series P_t^X (for X=S, A, M) will be investigated.

$$\vec{f}^X = (f_1^X, \ldots, f_i^X, \ldots, f_{B_x}^X)$$
$$\vec{b}^X = (b_1^X, \ldots, b_j^X, \ldots, b_{B_x+1}^X) \quad (1)$$
$$\vec{f}^X = |\{P_t^X | b_i^X < P_t^X \leq b_{i+1}^X\}|$$

$\vec{b}^X$ denote the respective interval borders.

The basic idea of downscaling $\vec{f}^A$ to $\vec{f}^S$ is that for days within a given interval of areal precipitation P_t^A different station values P_t^S may occur, depending on metrological characteristics like advective/convective rains, orography etc. For an illustration see Figure 2. Here each point stands for one day, on the x-axis the area precipitation value is displayed while the y-axis describes the respective precipitation at the station. The grid denotes exemplary bins, defined by $\vec{b}^A$ and $\vec{b}^S$. In case that the station data would reproduce exactly the areal average, all data points would lie on the grey diagonal line. Now the temporal overlap period between the P_t^A and P_t^S time series can be used to determine the respective transformation matrix $\mathbf{D}$ which describes the statistical relation between areal and station data:

$$\vec{f}^S = \mathbf{D}\vec{f}^A \quad (2)$$

$$D_{i,j} = |\{(P_t^A, P_t^S) | b_j^A < P_t^A \leq b_{j+1}^A \wedge b_i^S < P_t^S \leq b_{i+1}^S\}|/f_j^A$$

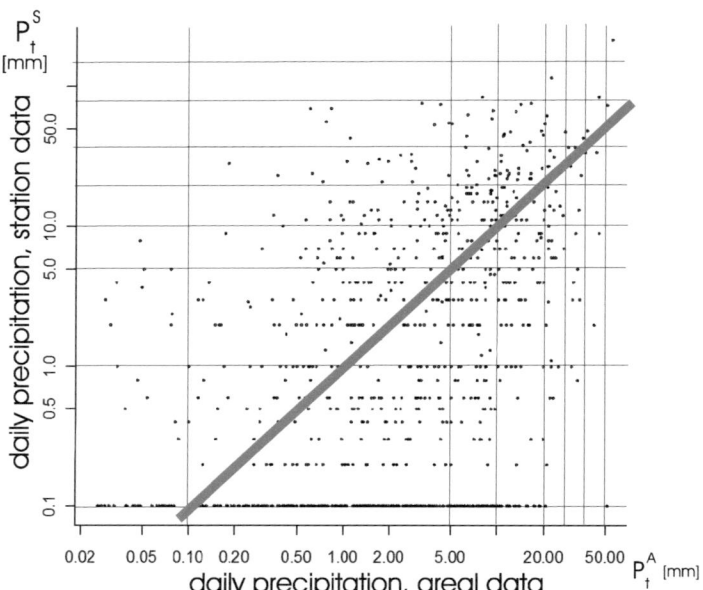

Figure 2: Relation of areal and station daily precipitation data
Source: own work, exemplary data from xDat and IMD

Applying **D** on model based projections of $\vec{f}^M$ will yield projections of the frequency distribution at the station level, and contain the model errors with respect to the reproduction of the current climate. To avoid this, we adjust the modelled distribution to the observed areal data by first scaling P_t^M to the observed maximum precipitation

$$P_t^{M'} = e \cdot P_t^M \ , \ e = \frac{max(P_t^A)}{max(P_t^M)} \tag{3}$$

and correct then for the remaining differences by the diagonal matrix **E**

$$E_{i,j} = f_i^A / f_i^{M'} \ , \ E_{i,j_{i \neq j}} = 0 \tag{4}$$

With this, the observed distribution of the station data is exactly reproduced for the current time period:

$$\vec{f}^S = \mathbf{DE}\vec{f}^{M'} \tag{5}$$

Sometimes the time span of the overlap between the daily time series P_t^S and P_t^A may be too short to make reliable statements on the frequency of rare events, but the whole time series at the station does allow for a distribution $\vec{f}^{S'}$ which is also adequate for more extreme events. To introduce this additional information into the downscaling transformation for future time slices a last correction is introduced.

$$C_{i,j} = f_i^{(S')}/f_i^S \ , \ C_{i,j_{i \neq j}} = 0 \tag{6}$$

So, using the observed precipitation time series, P_t^S and P_t^A, and the AOGCM output for the current climate, P_t^M, the transformation matrix $\mathbf{T} = \mathbf{CDE}$ and the scaling factor e are determined and can be applied to the future model outputs $^{\pm 2050}P_t^M$ and $^{\pm 2100}P_t^M$, yielding the model-error corrected and downscaled precipitation frequency distributions for the respective time slices, $^{\pm 2050}\vec{f}^{S'}$ and $^{\pm 2100}\vec{f}^{S'}$. In the following we summarise the procedure:

1. step: determine the frequency distribution $^{timeslice}\vec{f}^{M'}$ of $e \cdot {}^{timeslice}P_t^M$

2. step: apply the transformation matrix $^{timeslice}\vec{f}^{S'} = \mathbf{CDE}\ ^{timeslice}\vec{f}^{M'}$

2.2 Evaluation Algorithm for Heat Waves

Two variables are chosen to describe heat waves climatologically: (1) the number of days per year with minimum temperatures greater a given threshold and (2) the frequency distribution of the duration of heat waves, where the duration is defined as the number of consecutive days with the latter property. This is necessary because, e.g., in case the number of hot days doubles it makes a difference if these additional very hot days are consecutive or "interrupted" by some cooler ones generating some relief. To allow for a proper statistics in this first approach, we choose a relatively low threshold night temperature of 27 °C along with the definition of National Weather Service (2009). This is only a first approximation to assess the projected changes in the frequency of heat events following the more restrictive definition of the IMD. In the second project phase, we hope to be able to do estimations of more extreme events on the basis of the whole IPCC AR4 model set.

The downscaling problem of the daily minimum temperature as calculated by the AOGCMs, $_{min}T_t^M$, for the whole grid element to the situation in Hyderabad, $_{min}T_t^M$, is not as critical as for the much more spatially inhomogeneous precipitation data and the simple calibration of the areal AOGCM result on the urban station data is adequate.

Figure 3 compares the frequency distribution $_{min}T_t^S$ at the Begumpet station in Hyderabad (the green curve) with the distribution resulting from $_{min}T_t^M$ of the Echam5-run. The latter is separated into two 20-year time spans to show that this period is sufficient

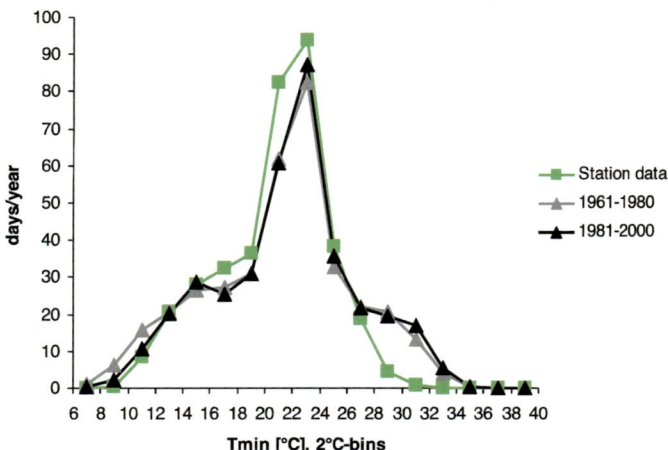

Figure 3: Frequency distribution for $_{min}T_t^S$ at the Begumpet station (green curve) and for $_{min}T_t^M$ of the Echam5-run, separated in two 20-year time spans (grey and black curves)

Source: own work, xDat and CMIP3 dataset

for determining this distribution (grey and black curves, almost identical). Comparing these with the green curve shows qualitatively good agreement ($_{min}T_t^S$ value for the maximum frequency is exactly reproduced), but in particular from 27 °C on the curves deviate. This makes clear that calibration is necessary.

There are different plausible possibilities to transform $_{min}T_t^M$ to approximate the green distribution (according to min/max, the average, the standard deviation etc.). We choose here, according to our aim, the number of days per year exceeding the threshold $T_c = 27\,°\mathrm{C}$. To do this in the most simple way, we shift T_c towards a (in this case) higher value $T_c{}'$ to obtain the expected number of warmer days for the current run. This value is then held fixed and used for the evaluation of the modelled future time slices.

3 Data Sources and Preparation

In the following section we describe the sources for the observational data (weather station and areal) as well as the model runs. Finally we document the necessary adaptations in temporal and spatial scale to make the different sources comparable.

3.1 Observational Data

The first kind of observational data needed are time series of daily precipitation, daily minimum and daily average temperature at a weather station in Hyderabad. The longest time series within the urban area of Hyderabad, at least as we know, is the weather station in Begumpet (former international airport), North of the lake Hussein Sagar. Here we could retrieve the almost complete time series for the period 1997 to 2007 from the xDat – PIK database. Additionally, for the same station we could retrieve the respective data for 1901–1970 from the KNMI-data base, which contains some more gaps but is still very valuable for determining statistical properties. Unfortunately, even the India Meteorological Department (IMD) National Data Centre is not able to provide data for the missing time period for any weather station within the urban area – they have only data available from agro-meteorological stations in the periphery of Hyderabad for scattered time periods which are not adequate to complete the picture due to the urban heat island effect which is—fortunately—reflected in the Begumpet data.

The second kind of observational data is on areal averages of daily precipitation in the larger area which includes Hyderabad. As the AOGCM results are of this kind, the observational analogue is important for downscaling and calibration. Here the IMD Climate Research Centre proved to be a valuable resource as they could provide areal daily precipitation data for India in a spatial resolution of $1° \times 1°$ for the time period 1951–2004 (an actualisation is under way). This data was generated by interpolation of station data and it has to be considered that the quality of this time series varies in time due to changing station density. An interesting possibility to crosscheck this purely "ground based" data set is the work of Kotteck and Rubel (2005) who combined satellite remote sensing data with station data to generate a global $1° \times 1°$ dataset of daily precipitation for the time period 1997–2005 (available at POSTEL).

3.2 AOGCM Model Runs

The AOGCM-results were taken from the CERA and the World Climate Research Program's (WCRP's) Coupled Model Inter-comparison Project phase 3 (CMIP3) multi-model dataset which provides, amongst others, the runs performed for the IPCC, AR4 process.

As we need daily data values for precipitation and minimum/average temperature, the usual CERA account occurs to be insufficient and one has to use the CMIP3 archive. Figure 4 gives an overview on the calculations performed by the 23 AOGCMs for AR4.

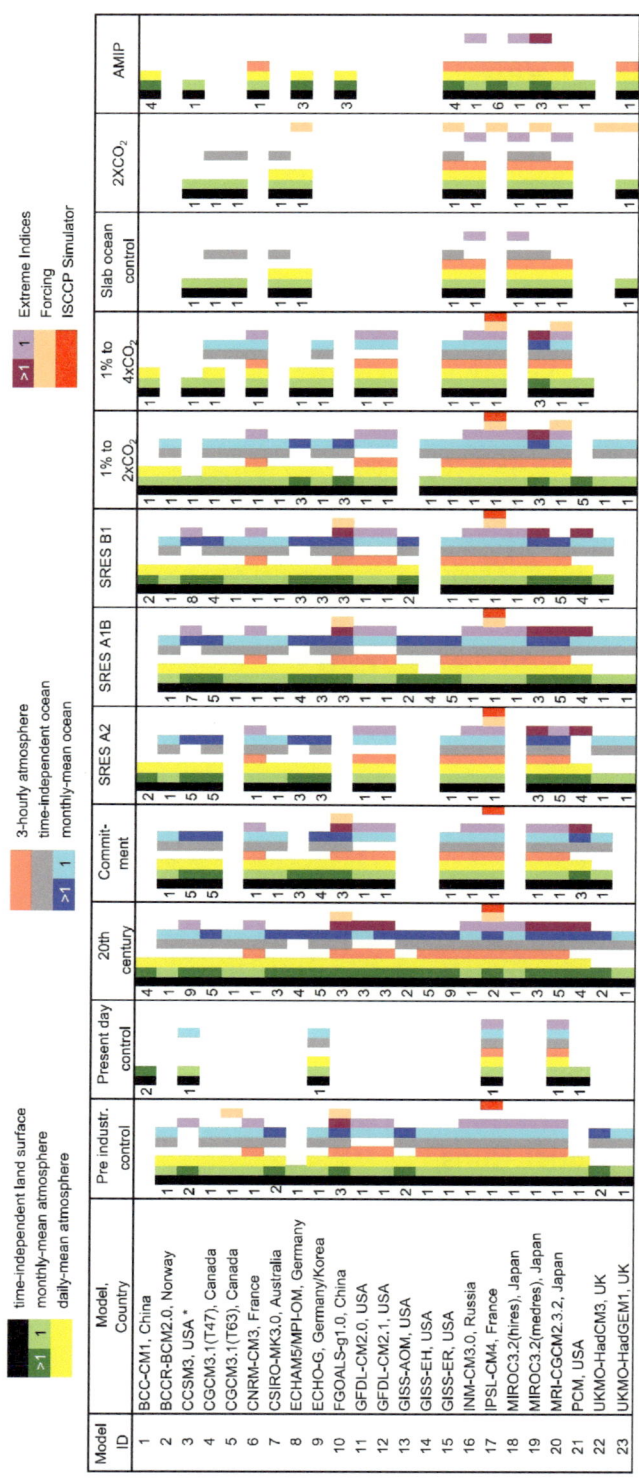

Figure 4: Runs provided by the different AOGCMs for IPCC-AR4
Source: IPCC, AR4 WG1, chapter 10, p 761 Meehl et al., 2007

As already mentioned, for this paper we chose the ECHAM 5 and the GFDL-CM2.1 models, the global SRES-emission scenarios A2 and B1 and therefore the runs 20C3M (i.e., the climate of the 20th Century experiment), SRESA2 (i.e., the SRES A2 experiment) and SRESB1 (i.e., the 550 ppm stabilisation experiment). All data can be retrieved in netcdf-raster format, including the exact geo-referencing of the data which has a resolution of about $2° \times 2°$.

3.3 Data Preparation

Besides the usual homogenisation of the temporal scales of daily time series (representation of leap years etc.), the different spatial scales of the areal data sets needed most attention (see Figure 5). Depending on the climate variable, the observed $1° \times 1°$ areal data was adapted to the AOGCM grid cell by either a weighted mean (temperature) or using only these $1° \times 1°$ grid cells which relevantly contribute to the AOGCM grid cell.

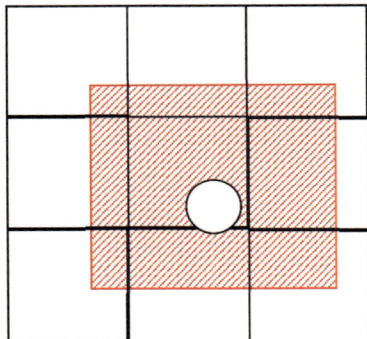

Figure 5: Schematic diagram showing the typical relation between the observed (black squares) and modelled (red square) areal data. The white circle characterises the urban area.

Source: own work

4 Results

The following sections show the results for Hyderabad and the Hyderabad region after applying the evaluation mechanisms (see Section 2) to the data sets described in Section 3.

4.1 Projections of the daily Precipitation frequency distribution

Figure 6 shows the current frequency distribution of daily precipitation at the Begumpet station. Frequency is characterised by the number of days in one year with a daily precipitation within the interval denoted on the x-axis. A number of 0.47 days/year means that (on average) each two years we can expect a daily precipitation event between 80 and 160 mm. The monsoon climate is reflected in (on average) 276 dry days per year.

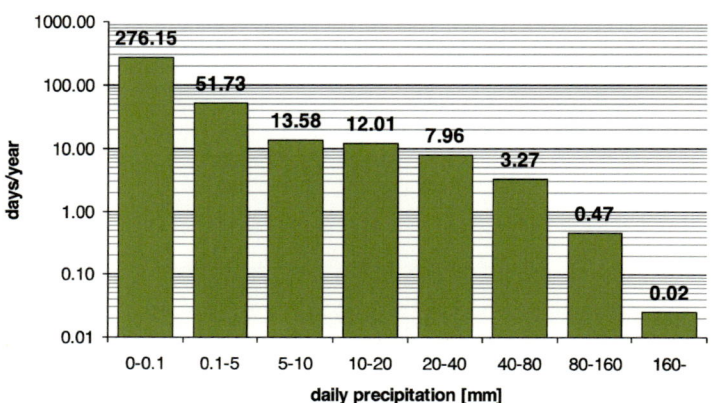

Figure 6: Frequency distribution of daily precipitation at the Begumpet station, current climate

Source: own work based on PIK-xDat database

After applying the evaluation algorithm for correction and downscaling (Section 2.1) to the chosen model runs of the Echam 5 and GFDL AOGCMs one gets the relative changes of the frequency for each interval shown in Figure 6. These results are summarised in Figure 7, depicting the situation for the two global emission scenarios A2 and B1 during the two future time slices. The two colours denote the two models, giving a first idea of uncertainty ranges which will be quantified more rigorously in the following project phase.

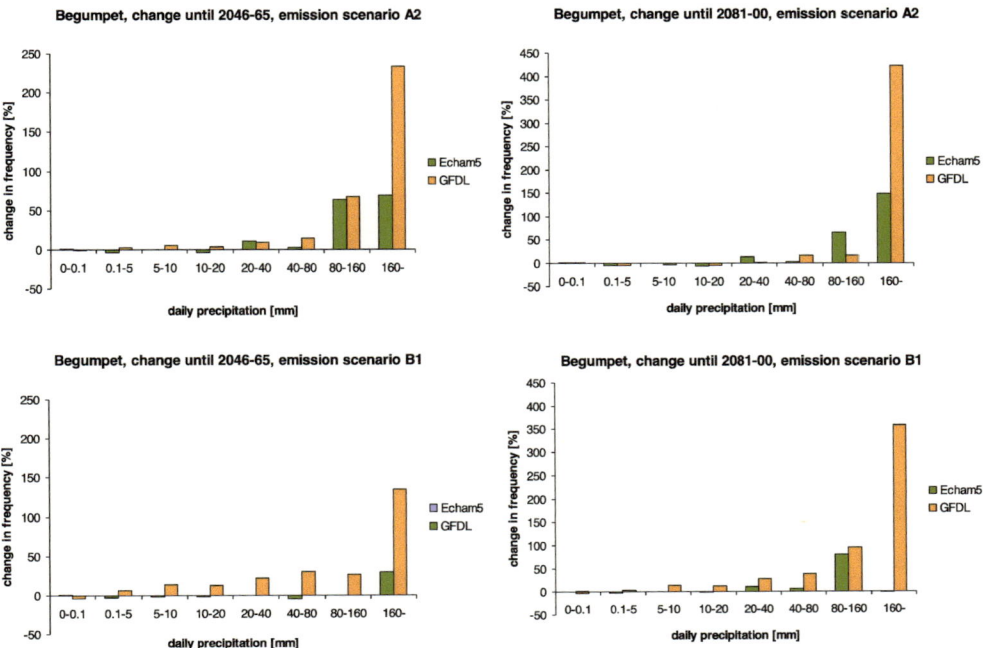

Figure 7: Projected relative changes in the frequency of daily precipitation events at the Begumpet station as calculated from two different AOGCMs and for two different global emission scenarios.

Source: own work (Section ref4-2009-chapter2-1), PIK-xDat database, IMD-database, KNMI-database, CERA-database

Qualitatively, all four diagrams show an increasing rate of change with higher precipitation values, in many cases the "160 mm +" -bin shows even an extreme increase up to 400 %! The GFDL-Model shows generally higher rates of change compared to the Echam 5 results. As the last two bins (> 80 mm) denote impact relevant precipitation amounts we additionally added up these two and got the following rates of problematic change in extreme precipitation events:

Table 1: Change in the occurrence of extreme daily precipitation (> 80 mm/day) compared with the current situation

	2045–2065	2081–2100
Scenario A2	70 ±6 %	53 ±17 %
Scenario B1	17 ±16 %	93 ±16 %

Source: own work

4.2 Projections of the annual Precipitation

As discussed in Section 2, for the areal perspective on annual precipitation (and temperature) we can use the rates of change resulting from the direct comparison of the outcome of the AOGCM runs.

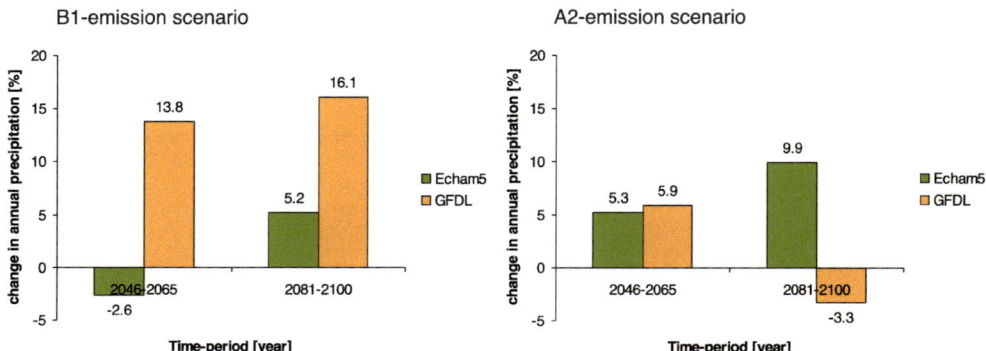

Figure 8: Projected relative changes in the amount of annual precipitation for the Hyderabad region as calculated from two different AOGCMs and two different global emission scenarios

Source: own work, CERA-database

The results are displayed in Figure 8, showing large differences between the chosen models, including the sign of change and the temporal trend within the 21st century. In general, the amount of potential reduction is lower than the possible increase, in B1 we even see the largest increase. This ambiguous situation resembles the result shown in Figure 1, so that we can anticipate that the full uncertainty analysis of the next phase will probably not change the picture for this variable.

4.3 Projections of heat waves

For projecting the future number of very hot nights per year, we apply the procedure explained in Section 2.2 to the AOGCM runs for the different emission scenarios and time slices. The frequency distributions derived directly from the runs are displayed in Figure 9. After the calibration we got the relative changes as displayed in Figure 10. The trend during the 21st century occurs to be monotonous and from 2050 on clearly stronger under the A2 scenario then under the B1 scenario. The number of hot nights is projected to increase by 500-600 % until 2100 under A2. This would mean in absolute terms an increase from 16 very hot nights per year to about 100! Under the emission reduction scenario B1, we would expect a more moderate increase towards about 50 days/year. A further important variable is the length of the heat waves. Figure 11 shows the frequency distribution of heat wave events with different length for the Begumpet station in Hyderabad. It tells us, for example, that we have on average three one-day heat waves per year, while a heat wave of six days occurs only once in two years. To get an idea about future change of more extreme events, we look for heat waves of one week or more. This occurs in the current climate about once in three years.

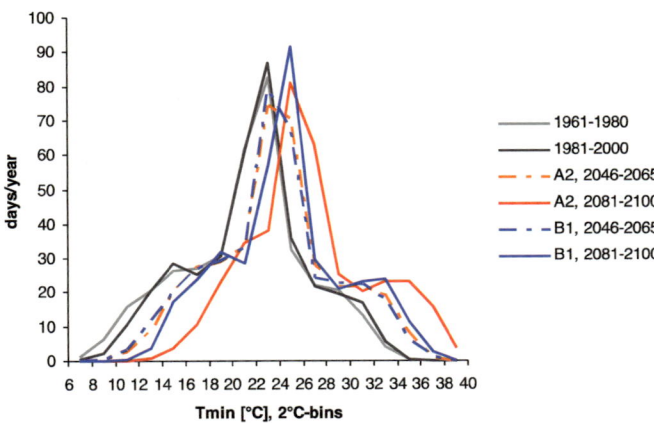

Figure 9: Frequency distribution for $_{min}T_t^M$ of the Echam 5-run, for different time slices and the emission scenarios A2 and B1
Source: own work, CMIP3 dataset

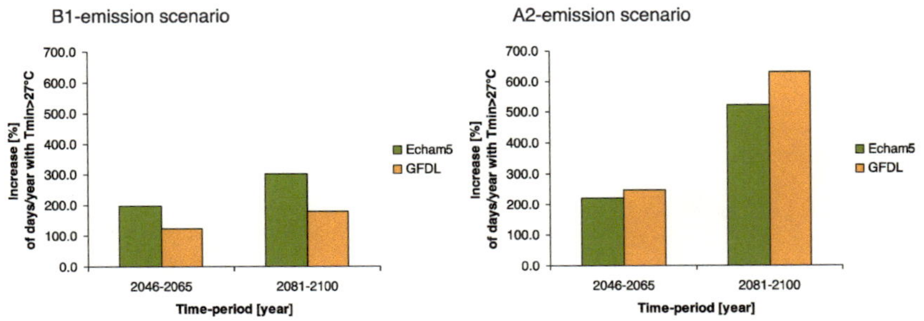

Figure 10: Projected relative changes in the number of day per year with $_{min}T_t^M$ warmer than 27 °C in Hyderabad for Echam 5-and GFDL-runs under two different global emission scenarios.
Source: own work (Section 2.2), PIK-xDat database, CMIP3 dataset

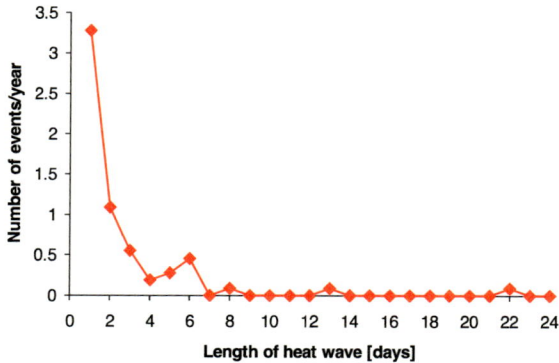

Figure 11: Frequency distribution of the length of heat waves for the current climate in Hyderabad, Begumpet station.

Source: own work, PIK-xDat database

In Figure 12 the respective projections are displayed. The pattern resembles roughly the expected change in the number of very hot nights. Until 2050 we have to prepare for seeing this event each year, relatively independent from the emission scenario. Under B1 this situation will persist until 2100. Under A2 the model results are more ambiguous for this time horizon, ranging from a 200 to a 500 % increase.

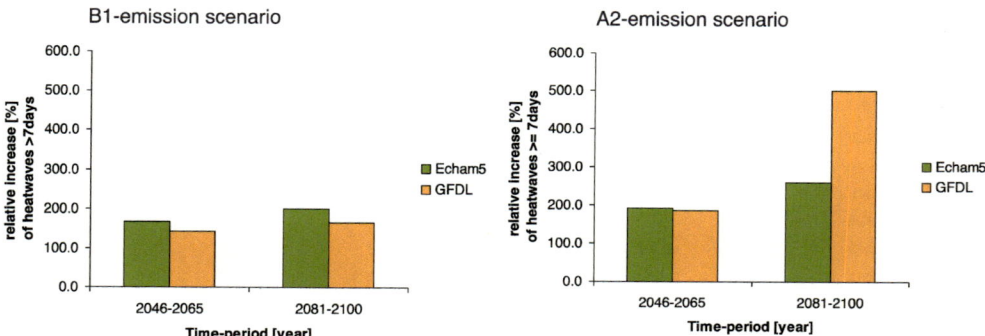

Figure 12: Projected relative changes in the number of heat wave events of one week or longer in Hyderabad for Echam 5-and GFDL-runs under two different global emission scenarios.

Source: own work (Section 2.2), PIK-xDat database, CMIP3 dataset

4.4 Projections of the annual average temperature

These projections as direct evaluations of the AOGCM runs, are displayed in Figure 13 and follow closely the pattern of the evaluation of the number of hot nights (see Figure 10). The projected 5 °C increase till 2100 under A2 would massively influence the natural water balance towards dryness, even under a moderate increase in annual precipitation.

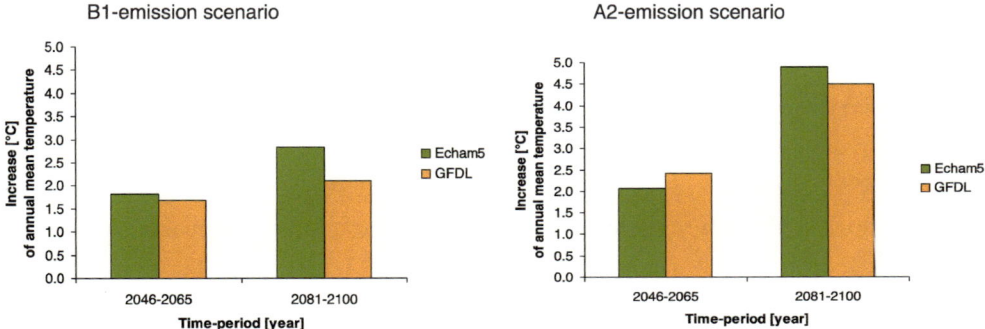

Figure 13: Projected relative changes in annual average temperature in the Hyderabad area for Echam5- and GFDL-runs under two different global emission scenarios.

Source: own work, CMIP3 dataset

5 Conclusions and Policy Implications

The paper develops adequate evaluation algorithms for applying AOGCM (atmospheric - oceanic general circulation model) results to the Hyderabad situation. The AOGCM results we use here were generated for the IPCC AR4 process and we refer to runs for a high (A2) and a low (B1) global CO_2-emission scenario. The considered time slices are 1981-2000 (reference climate), 2046–2065 and 2081–2100. We evaluate these model runs to obtain projections of the four most impact-relevant climatic characteristics for Hyderabad: the frequency distribution of daily precipitation (important, e.g, for urban flooding), the total annual precipitation (e.g. for urban water supply), the probability and duration of heat waves (e.g. for human health) and the annual mean temperature (e.g. for urban agriculture). Although the full fledged uncertainty analysis will be a task of the following project phase, we can already give first ranges for quantitative best-guess projections, resulting in a much differentiated picture. For daily precipitation greater than 80 mm/day in Hyderabad (medium to very high impact intensity with respect to damages), we calculated an increase in frequency of about 70 % (± 6) until 2050 for the high emission scenario. This worsened situation tends then to be relatively stable until 2100. For the low emission scenario, the opposite is the case: here we expect a low increase (17 % ± 15) until 2050 followed by a period of rapid increase (93 % ± 15, compared to 2000). In case this result will be finally corroborated by the full uncertainty analysis a first important policy implication is that a massive reduction of the global CO_2 emissions (B1 instead of A2) will certainly "buy" some time for Hyderabad to adapt to more intensive rainfalls (in the second instead of the first half of the 21st century) but will not spare the city to prepare for about a doubling of strong rain events. This is even worsened by the result that the subset of extreme events greater 160 mm/day (like, e.g., in August 2000) will increase over-proportionally. This shows the priority of adapting to pluvial flooding by improving the drainage infrastructure and/or making the urban functions less sensitive regarding these floodings. Well based quantitative uncertainty estimates will be given in next phase, but a quadruplicating of the frequency of very strong rain events is within the possible results (i.e. a 40 year event will occur each ten years–on average). This results go along with the trend, in the time period 1951–2000, for increased extreme rain events over India as mentioned in Goswami et al. (2006) and the trend over the past century in different extreme indices (Sen Roy and Balling 2004).

For the total annual precipitation in the region around Hyderabad (about 100 km in radius), we have to expect changes between -4 % to +17 %; the influence of the differences in the AOGCMs is larger than the influence of the global emission scenarios. So our actual best guess is that Hyderabad has to prepare for a change in average annual precipitation of about ± 20 % and that it is neither possible to predict the sign nor the exact amount–this may either be caused by still existing shortcomings of the AOGCMs (in this case the precision will improve) or by principal properties of the monsoon system (in this case this uncertainty is intrinsic and not to avoid). Again, this situation has important consequences for planning the future water supply of Hyderabad. The large uncertainty in future annual rainfall (which may be systemic) demands for a flexible planning strategy for the water supply of Hyderabad–in general the sensitivity should be minimised by avoiding demand increase where ever possible.

For the different aspects of heat waves this analysis can give again some quantitative estimates. Here the increase of the first indicator, denoting the average number of days per year with night temperatures $> 27\,°C$ will approximately triple until 2050. This is relatively independent from the emission scenario, but in 2100, A2-emissions will lead to an increase of almost 560 % (± 50) of the current number while under B1-emissions we expect only 240 % (± 50). The frequency of heat waves longer than one week will double to triple until 2050 and increase further until 2100. Here the situation differs from the pluvial flooding case which is already a severe problem under the current climate in Hyderabad that will become worse under climate change. Heat waves, on the other hand, are presently not regarded as a priority problem in Hyderabad but will become massive in the future thereby introducing a new problem dimension. This constitutes a particular challenge for planned adaptation as neither public awareness nor experiences in dealing with the problem exist.

The mean annual temperature will develop monotonously in time and with a stronger trend in the high emission scenario up to +5 °C. This high value of +5 °C for the A2 scenario in 2100 would definitely alter the natural water balance towards increased dryness, even under a (very uncertain) increase in total rainfall. Here the main policy implication is that a global emission reduction along the B1 scenario would clearly ease the adaptation pressure.

To summarise, this paper showed different behaviours of the relevant climate variables in response to global emission scenarios and allows thereby identifying adaptation necessities, which are more or less dependent on the future development of the global greenhouse gas emissions.

References

Literature

Goswami, B.N; Venugopal, V.; Sengupta, D.; Madhusoodanan, M.S. and Xavier, P. K. (2006). Increasing Trend of Extreme Rain Events Over India in a Warming Environment. Science 314: 1442–1445.

Hollweg, H.-D.; Böhm, U.; Fast, I.; Hennemuth, B.; Keuler, K.; Keup-Thiel, E.; Lautenschlager, M.; Legutke, S.; Radtke, K.; Rockel, B.; Schubert, M.; Will, A.; Woldt, M. and Wunram, C. (2008). Ensemble Simulations over Europe with the Regional Climate Model CLM forced with IPCC AR4 Global Scenarios. Technical Report SGA-ZMAW.

Reckien D, Lüdeke M, Reusswig F, Kit O, Meyer-Ohlendorf L, Budde M (2011): Hyderabad, India, infrastructure adaptation planning. In Rosenzweig C, Solecki WD, Hammer SA, Mehrotra S: Climate Change and Cities – First Assessment Report of the Urban Climate Change Research Network, Cambridge: Cambridge University Press, ISBN 978-1-107-00420-7: 152–154.

Kottek, Markus and Rubel, Franz 2005. CSP – Algorithm Theoretical Basis Document (ATBD): WP 8318 – CSP Precipitation. IP geoland, EC Proposal Reference No. FP6-502871.

McMichael, A.J.; Campbell-Lendrum, D.H.; Corvalan, C.F.; Ebi, K.L.; Githeko, A.K.; Scheraga, J.D. and Woodward, A. (eds.) (2003). Climate Change and human Health: Risks and Responses, World Health Organization, Geneva 333 pp.

Meehl, G.A.; Stocker, T.F.; Collins, W.D.; Friedlingstein, P.; Gaye, A.T.; Gregory, J.M.; Kitoh, A.; Knutti, R.; Murphy, J.M.; Noda, A.; Raper, S.C.B.; Watterson, I.G.; Weaver, A.J. and Zhao, Z.-C. (2007). Global Climate Projections. In: Solomon, S.; Qin, D.; Manning, M.; Chen, Z.; Marquis, M.; Averyt, K.B.; Tignor, M. and Miller, H.L. (eds.). Climate Change 2007: The Physical Science Basis. Contribution of Working Group I to the Fourth Assessment Report of the Intergovernmental Panel on Climate Change. Cambridge University Press, Cambridge, United Kingdom and New York, NY, USA.

Nakićenović, N. and R. Swart (eds.). 2000. *Special Report on Emissions Scenarios: A Special Report of Working Group III of the Intergovernmental Panel on Climate Change.* Cambridge University Press, Cambridge, United Kingdom and New York, NY, USA, 599 pp.

Orlowsky, B., F.-W. Gerstengarbe and P.C. Werner. 2008. "A resampling scheme for regional climate simulations and its performance compared to a dynamical RCM." *Theor. Appl. Climatol.* 92(3-4): 209–223.

Revi, A. 2007. "Climate change risk: an adaptation and mitigation agenda for Indian cities." Paper held at the *Global Urban Summit* organised by the Rockefeller Foundation in Bellagio, 8-13th July 2007.

Rupa Kumar, K., A.K. Sahai, K. Krishna Kumar, S.K. Patwardhan, P.K. Mishra, J.V. Revadekar, K. Kamala and G.B. Pant. 2006. "High-Resolution Climate Change Scenarios for India for the 21st Century. Indian Institute of Tropical Meteorology, Pune." *Current Science* 90(3): 334–345.

Satterthwaithe, D., S. Huq, M. Pelling, H. Reid and P. R. Lankao. 2007. "Adapting to Climate Change: The possibilities and constraints in low- and middle-income nations." In *Human Settlements Discussion Paper Series: Theme: Climate Change and Cities 1*.

Sen Roy, S. and R.C. Balling. 2004. "Trends in extreme daily precipitation indices in India." *International Journal of Climatology* 24: 257–466.

Tebaldi, C., K. Hayhoe, J.M. Arblaster, G. and Meehl. 2006. "Going to extremes. An intercomparison of model-simulated historical and future changes in extreme events." *Climatic Change* 79: 185–211.

National Weather Service; Silver Spring, Maryland, USA (25.05.2009). www.nws.noaa.gov/glossary/index.php?word=heat+advisory.

Data Source

xDat (eXtensible Database Access Tool). Potsdam Institute for Climate Impact Research. www.pik-potsdam.de/institute/organization/scientific-departments/data-computation/sdm/tools/xdat/

KNMI (Koninklijk Nederlands Meteorologisch Instituut)-Database, Wilhelminalaan 10, 3732 GK De Bilt, Netherlands

IMD (India Meteorological Department)- National Data Centre, National Climate Centre, Shivajinagar, Pune – 411005, India

POSTEL (Pôle d´Observation des Surfaces Continentales par Telédétection, MEDIAS-FRANCE) - CNES - Bpi 2102; 18 avenue Edouard Belin; 31401 TOULOUSE Cedex 9, FRANCE; http://postel.mediasfrance.org/en/DOWNLOAD/Biogeophysical-Products/

CERA / IPCC DDC (Climate and Environmental Retrieving and Archiving, DKRZ), http://cera-www.dkrz.de/CERA/index.html

WCMIP3 World Climate Research Program's (WCRP's) Coupled Model Inter-comparison Project phase 3 (CMIP3) multi-model dataset. We acknowledge the modelling groups for making their model output available for analysis, the Program for Climate Model Diagnosis and Inter-comparison (PCMDI) for collecting and archiving this data, and the WCRP's Working Group on Coupled Modelling (WGCM) for organising the model data analysis activity. The WCRP CMIP3 multi-model dataset is supported by the Office of Science, U.S. Department of Energy.